Published by Ice House Books

TM © 2020 IFLSCIENCE Limited. All rights reserved.

Written and Designed by Smart Design Studio

Photography Shutterstock.com – for individual credits see back page

Ice House Books is an imprint of Half Moon Bay Limited
The Ice House, 124 Walcot Street, Bath, BA1 5BG
www.icehousebooks.co.uk

ISBN 978-1-912867-67-7

Printed in China

IFLSCIENCE!

THE EARTH IS FLAT & OTHER STORIES

ICE HOUSE BOOKS

THE EARTH IS FLAT

NASA IS LYING TO US,
THE WORLD ISN'T SPHERICAL AT ALL...

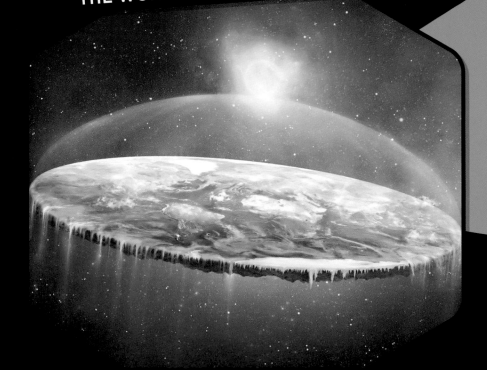

NOT TRUE!

Since the days of Aristotle, scientists have known the Earth is round.

And it's easy to prove...

TIME ZONES: When it's lunchtime in NYC, it's midnight in Beijing. Because the world is round, and the sun can't shine on all of it at once. **Duh.**

PLANES: You can fly all around the world and not reach the edge.

OTHER PLANETS: Are round. So why would the Earth be any different?

SORRY, FLAT-EARTHERS, MYTH BUSTED!

THE GREAT WALL OF CHINA

...IS NOT VISIBLE FROM SPACE!

Have you ever been to space and seen the Great Wall of China? **No?**

Well neither have the astronauts.

TOTAL MYTH

NO SIGNS OF CIVILISED LIFE ON THIS ONE...

THE AGE OF THE DINOSAURS

A MASS EXTINCTION EVENT SEPARATES THE EXISTENCE OF DINOSAURS AND HUMANKIND, BUT STILL...

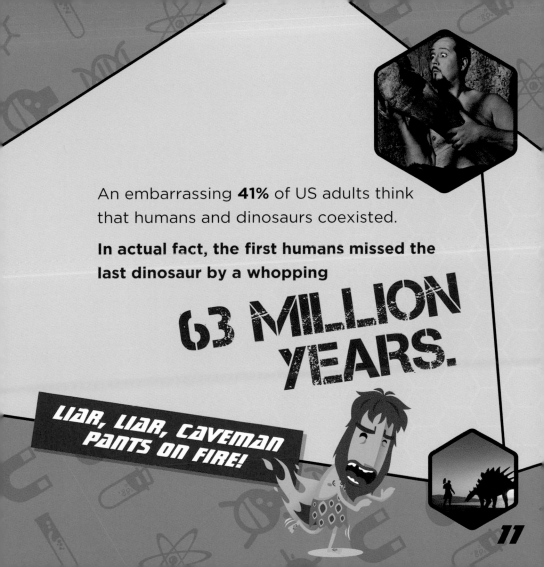

An embarrassing **41%** of US adults think that humans and dinosaurs coexisted.

In actual fact, the first humans missed the last dinosaur by a whopping

63 MILLION YEARS.

LIAR, LIAR, CAVEMAN PANTS ON FIRE!

RUDOLF HESS

THE NAZI GENERAL WAS REPLACED WITH A DOPPELGANGER BEFORE STANDING TRIAL.

After years of speculation, scientists were able to prove once and for all that Nazi war criminal, Rudolf Hess, really did get his comeuppance.

A vial of Hess's blood, and DNA extracted from a living descendant, was able to prove beyond doubt that the right man was IMPRISONED.

JUSTICE WAS SERVED!

STRANGER THAN STRING THEORY

DID STEPHEN HAWKING DIE IN 1985 AND GET REPLACED WITH A NASA-CONTROLLED LOOKALIKE?

THE THEORY: For the last 33 years of his 'life' Stephen Hawking was actually a lookalike actor.

THE EVIDENCE: In 1985 Hawking had brown hair, but in the most recent pictures of him his hair was grey **(we s**t you not, this is an actual argument)**.

THE CONCLUSION: People with motor neurone disease have varying life expectancies. Hawking really did beat the odds and live to the grand old age of 76!

THE UNIVERSE IS EXPANDING ... AND SO ARE THE CONSPIRACY THEORIES!

MOON LANDINGS

NASA LIED – APOLLO 11 NEVER LANDED ON THE MOON.

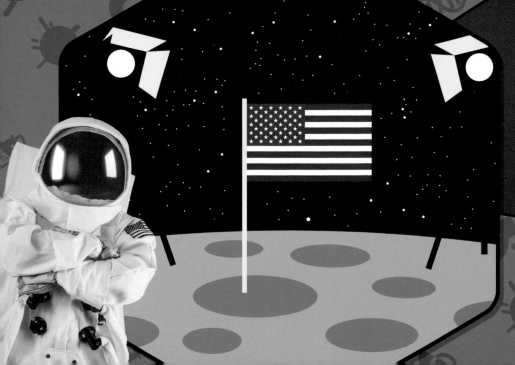

LET'S DEBUNK A FEW OF THE MOST COMMON ARGUMENTS:

Why are there no stars in the Moon landing pictures?	The stars are there, they're just too faint to be seen!
There was no flame from the rocket as it brought the astronauts back to Earth.	The fuel used produces no visible flame!
There's no atmosphere in space, so how can the US flag wave?	Because a flag can still wave in a vacuum!

ONE SMALL STEP FOR MAN, ONE GIANT LEAP FOR MANKIND?

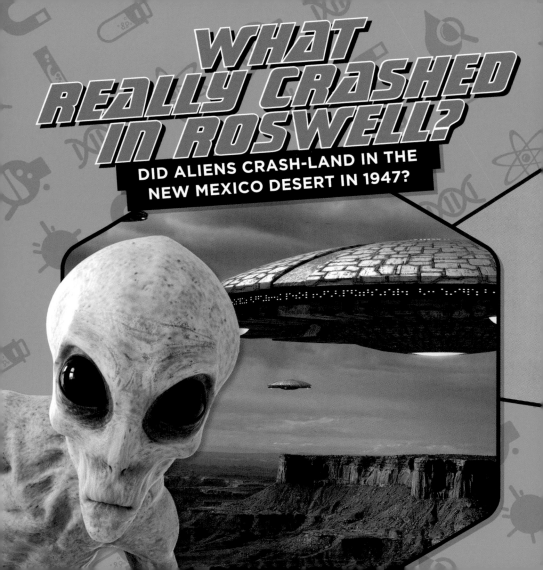

WHAT REALLY CRASHED IN ROSWELL?

DID ALIENS CRASH-LAND IN THE NEW MEXICO DESERT IN 1947?

Did something crash? **YES**

Did the government try to cover it up? **YES**

Was it an alien spacecraft? **ERR, NOPE.**

So, what was it? The truth finally came out in 1997 – the US Air Force was **testing a top-secret Cold War satellite**.

FACT: The US military was also crash-testing dummies from high altitudes around the same time.

19

GLOBAL WARMING

...IS A MYTH.

GLOBAL WARMING MYTH

- **The Earth's average temperature** has **risen** by **1.4°F** (0.8°C) since 1880

- **Arctic ice** is **thawing**

- The patterns of **migrating birds** are **changing**

- **Hurricane 'season'** is getting **earlier** each year

- **Flowers** are blooming **earlier** each year

CLIMATE CHANGE DENIAL IS THE REAL CONSPIRACY THEORY!

BIGFOOT IS REAL

...OR SO THE INDIAN ARMY WOULD HAVE YOU BELIEVE.

In April 2019 a tweet was posted to the official Twitter account of the Indian Army:

'For the first time, an **#IndianArmy** Mountaineering Expedition Team has sighted Mysterious Footprints of mythical beast 'Yeti' measuring 32x15 inches close to Makalu Base Camp on 09 April 2019.'

ACCOMPANYING THE TWEET WERE PHOTOS OF GIANT FOOTPRINTS.

Yeti? Mountain bear? Or someone with a wicked sense of humour?

THE SCIENTIFIC NAME FOR BIGFOOT = CRYPTOMUNDO

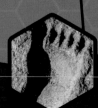

CHILDHOOD VACCINES

UP TO 20% OF THE POPULATION BELIEVES THAT VACCINATIONS CAUSE AUTISM.

This is scientifically

NOT TRUE!

A government-funded study confirms that there is **NO LINK** between vaccinations and autism.

Vaccinations protect us and others from terrible diseases.

FACT: Smallpox used to kill thousands but is now extinct because people vaccinated against it.

FLUORIDE in WATER

WHY DOES THE GOVERNMENT ADD FLUORIDE TO OUR DRINKING WATER?

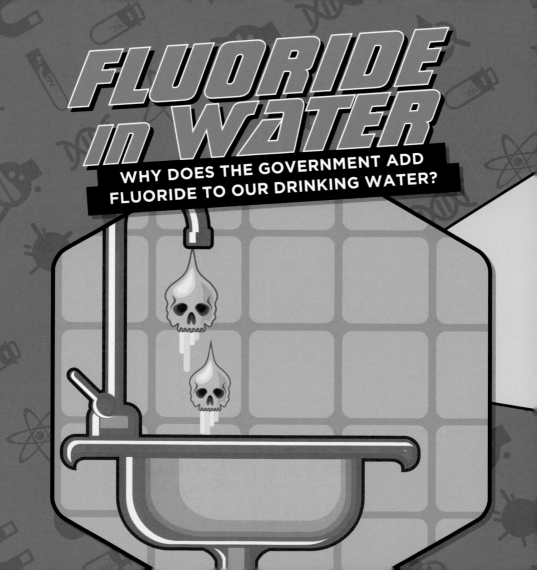

Fluoride prevents us getting cavities in our teeth, and that's why it's added to water.

However, conspiracy theorists claim that the **fluoride is a tranquiliser**, and the government is secretly **creating a zombie nation** by making us drink it.

Fluoride is a by-product of aluminium manufacturing. Is the government keeping big corporations happy by disposing of their chemical waste?

27

BUSTED!

We actually have close to

20 SENSES,

including **balance**, **pain**, **movement**, **hunger** and **thirst**....

SCIENCE DISPELS THE MYTH ONCE AGAIN!

JFK ASSASSINATION

EVIDENCE TAMPERING, A SECOND SHOOTER, CIA INVOLVEMENT ... THE STORIES AROUND PRESIDENT KENNEDY'S DEATH ARE NUMEROUS.

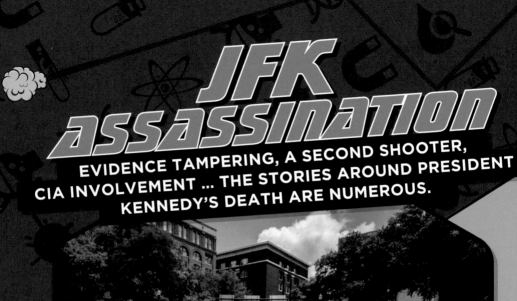

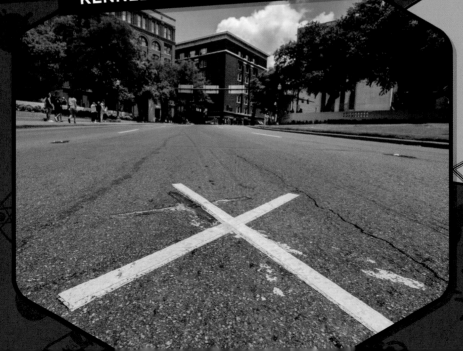

But science has successfully debunked **one popular theory:** one bullet couldn't have killed the President AND injured Governor John Connelly.

IN FACT, the trajectory of the bullet fired could plausibly have passed through Kennedy, killing him, and then hit the Governor.

IT'S ALL ABOUT THE GEOMETRY...

COUNTY MARINE
ARGED IN SLAY

CHEMTRAILS

ARE GOVERNMENT AEROPLANES LEAVING CHEMICAL TRAILS IN THE SKY?

Chemtrails

NO

Take off the tin hat and sit down ... no, the government is not trying to poison the air we breathe by flying planes with nefarious chemicals trailing out the back of them.

So why do some planes' trails stay in the sky longer than others? Because of differing atmospheric conditions, obviously!

STRIPES IN THE SKY!

KURT COBAIN'S SUICIDE

...WASN'T SUICIDE AT ALL.

THE THEORY:

Cobain had three times the lethal level of heroin in his system when he died, so how could he have pulled the trigger of a gun?

THE FACTS:

The Nirvana frontman was a long-time addict, with a very high tolerance level for heroin. Regular users also build up drug by-products in their blood, so lab results may be misleading.

PEOPLE ALSO ARGUE THAT COBAIN'S SUICIDE NOTE WAS FORGED!

BERMUDA TRIANGLE

FROM WHOSE WATERS NO TRAVELLER RETURNS...

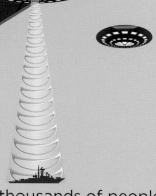

Have thousands of people really vanished without a trace in the Bermuda Triangle?

NO, DON'T BE ABSURD.

Most reports were either inaccurate or embellished.

SHIPS AND PLANES PASS THROUGH THE 'DEVIL'S TRIANGLE' EVERY DAY!

BLACK KNIGHT SATELLITE

AN ALIEN SPACECRAFT IS CONSTANTLY ORBITING THE EARTH...

...and NASA is covering it up!

SORRY... another wild story that science can put to bed.

The 'satellite' is actually a space blanket that was lost during an Extravehicular activity mission.

THEY'RE WATCHING US...

39

THE MOON ISN'T REAL!

IT'S REALLY A GIANT HOLOGRAM BEING BEAMED INTO THE SKY!

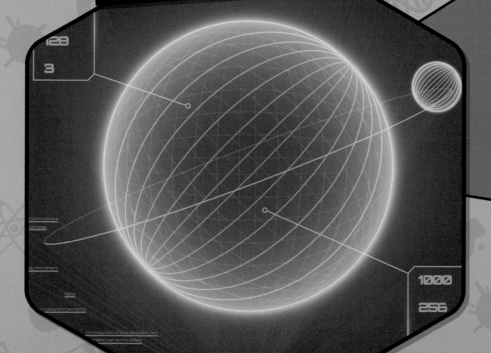

And moon rocks aren't real...
the tides control themselves...
lunar eclipses are a puppet show...

**Do we even need to dignify this
whack-attack c**p with a debunking?**

NO. JUST ... NO.

FACT: Only 12 people have ever stepped foot on the Moon.

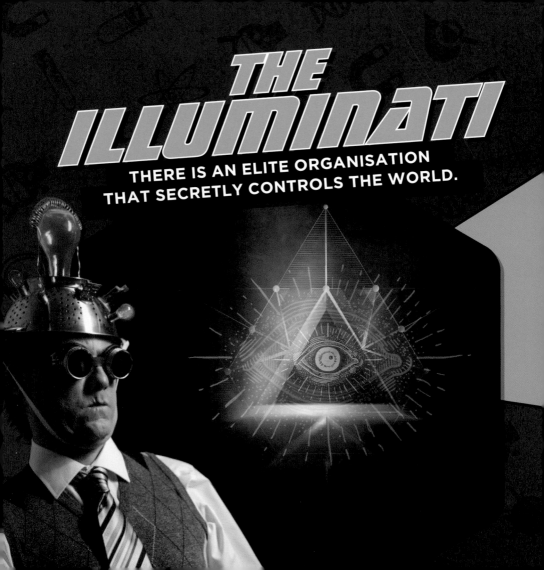

THE ILLUMINATI

THERE IS AN ELITE ORGANISATION THAT SECRETLY CONTROLS THE WORLD.

Scientists, **politicians**, **faith leaders**, even pop stars are all accused of being members of the highly secretive and highly unlikely **Illuminati organisation.**

THE EVIDENCE? Well, there isn't much really. However, people like to blame anything from the French Revolution to alien overlords on the Illuminati.

TRUE? Probably not. But the theory has spawned some awesome movies, at least.

43

PLANET X

SCIENTISTS ARE KEEPING THE EXISTENCE OF A 9TH PLANET (NO, NOT PLUTO)**, SECRET...**

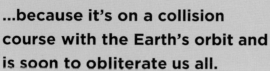

...because it's on a collision course with the Earth's orbit and is soon to obliterate us all.

Let out the breath you're holding ...
we can all sleep soundly in our beds tonight.

Countless amateur and professional stargazers have looked for this elusive planet and there's nothing there!

TIME FOR THAT ARMAGEDDON BUNKER?

EINSTEIN FLUNKED MATHS

IS THIS REALLY TRUE?

NO, IT'S NOT.

Sorry to burst your bubble –

Einstein really **WAS** a maths genius.

EINSTEIN DID ONCE FAIL A SCHOOL ENTRANCE EXAM, THOUGH!

$E=mc^2$

CIA USING MIND CONTROL

WHEN THE CONSPIRACY IS TRUE...

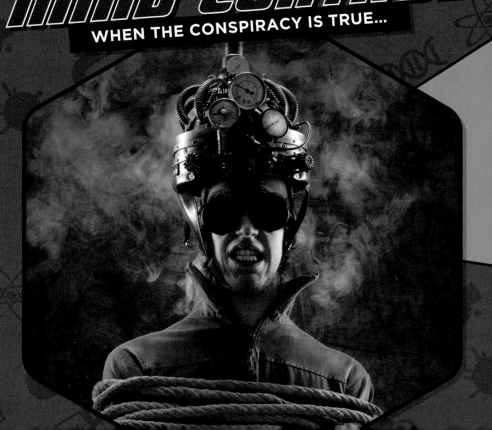

In the 1970s the US government declassified papers, which revealed that the secret service really had been dabbling in **mind control**, **electric shock therapy** and **psychological torture**.

THE REASON?

To develop techniques to use against their enemies.

150 human experiments took place between 1953 and 1964!

But as recently as the 1960s the advertising industry claimed that it was good for us.

Advertising companies paid a lot in taxes, and were powerful political lobbyists, and the government didn't say a word....

Smoking increases your risk of stroke, emphysema, infertility, and a s**t load of cancers.

SUGAR

GLOBAL COMPANIES ARE FUNDING RESEARCH TO SPIN THE FACTS...

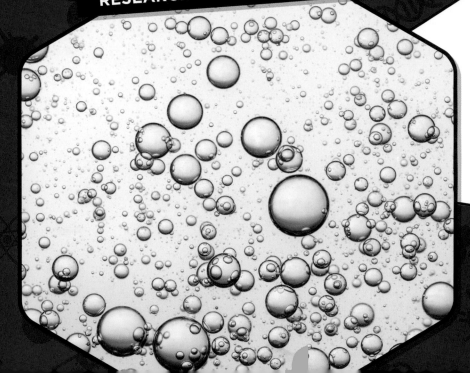

This conspiracy is

TRUE.

Some soft-drink companies have funded studies trying to link weight loss solely with exercise, in the hope that people will think drinking their product won't make them **gain weight.**

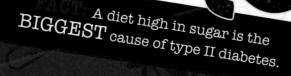

FACT: A diet high in sugar is the BIGGEST cause of type II diabetes.

NAZIS ON THE PAYROLL

DID THE US GOVERNMENT REALLY EMPLOY NAZI SCIENTISTS AFTER WW2?

YES

YES, THEY DID.

Over 1,600 Nazi scientists were put to work for the American Government after the end of the Second World War.

One scientist was even involved with the Moon landings!

THE TRUTH CAN BE STRANGER THAN FICTION...

REPTOIDS

GIANT LIZARD CREATURES FROM THE CENTRE OF THE EARTH ARE HIDING AMONGST US...

Not only that, but they are world leaders, members of the Royal Family and people of power and influence.

SCIENCE FICTION OR TERRIBLE TRUTH?

Do you really think the Queen of England rips off her mask every night to reveal her malevolent reptilian true form?

NO? Neither do we.

IT'S ALL IN THE EYES...

STILL DON'T BELIEVE US?

ACCORDING TO SCIENCE, CERTAIN PEOPLE BELIEVE CONSPIRACY THEORIES...

A recent study has shown that the lower the probability of an event, the more people embrace conspiratorial explanations.

Whether you question the `'TRUTH'` **or not depends on your understanding of probabilities, dislike of uncertainty, and desire to find an explanation for unlikely events.**

ARE YOU A BELIEVER?

PHOTO CREDITS